REPAIRING THE DAMAGE

Fires *and* Floods

David Lambert

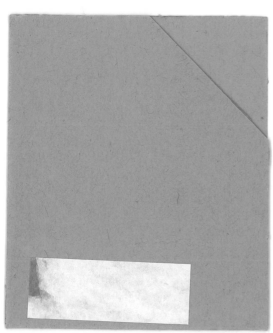

Cloverleaf
An imprint of Evans Brothers Limited

Cloverleaf is an imprint of Evans Brothers Limited

Evans Brothers Limited
2A Portman Mansions
Chiltern Street
London W1M 1LE

First published 1992

Typeset by Fleetlines Typesetters, Southend-on-Sea
Printed in Spain by GRAFO, S.A. – Bilbao

ISBN 0 237 51209 2

Acknowledgements

Series editor: Su Swallow
Editor: Nicola Barber
Design: Neil Sayer
Picture research: Elaine Finch
Production: Jenny Mulvanny
Maps and illustrations: Hardlines, Charlbury

For their help and information the publishers would like to
thank the following: Robert Wray of Eternit UK Limited;
Ministry of Agriculture, Fisheries and Food; Health and Safety
Executive.

For permission to reproduce copyright material the author and
publishers gratefully acknowledge the following:

Cover: (top) Daphne Kinzler, Frank Lane Picture Agency,
(bottom) Martin Bond, Science Photo Library
Title page (Khartoum, Sudan) C Hires, Frank Spooner Pictures
p4 (top) Michael Holford, British Museum, (bottom) Tweedie,
ECOSCENE **p5** (top) David C Houston, Bruce Coleman Ltd,
(bottom) Erwin & Peggy Bauer, Bruce Coleman Ltd **p7** John
Topham, Bruce Coleman Ltd **p8** David E Rowley, Planet Earth
Pictures, (inset) Frank Lane Picture Agency **p9** Daniel Dutka,
Rex Features/Sipa-Press, (inset) Eric Tschach, Rex Features/Sipa-
Press **p10** Jeff Foott, Bruce Coleman Ltd **p11** (left) NASA,
Science Photo Library, (right) Jesco von Puttkamer, Hutchison
Library **p12** Bolcina-Senepart, Rex Features/Sipa-Press **p13** e t
archive, (inset) Mary Evans Picture Library **p14** (top) Mary
Evans Picture Library, (bottom) Bernard Gerard, Hutchison
Library **p15** (left) Hulton Picture Company, (right) ZEFA **p16**
(top left) e t archive, (top right) Rex Features, (bottom) A de
Menil, Science Photo Library **p17** and **p19** Crown Copyright,
reproduced by permission of the Buildings Research
Establishment and the Controller of HMSO **p20** Mary Evans
Picture Library **p21** (top) Rex Features, (bottom) Eurotunnel,
Q A Photos Ltd **p22** Civil Aviation Authority **p23** P Uhl, ZEFA
p24 Tampyx, Frank Spooner Pictures **p25** Michael McKinnon,
Planet Earth Pictures **p27** (top) Italian State Tourist Office,
(bottom) Associated Press **p28** Moore & Moore, Robert
Harding Picture Library **p30** Steve McCurry, Magnum Photos
p31 (top) Barbara Klass, Panos Pictures, (middle) Heldur Jaan
Netocny, Panos Pictures, (bottom) Bartholomew/Liais, Frank
Spooner Pictures **p34** eyewitness account taken from *The
Greatest Disasters of the Twentieth Century* © Marshall
Cavendish, Hulton Picture Company **p35** Landform Slides **p36**
Philip Jones-Griffiths, Magnum Photos **p37** Popperfoto, (inset)
Anthony Cooper/ECOSCENE **p38** Orion Press/ZEFA **p39** Kotoh,
ZEFA, (inset) Bruno Barbey, Magnum Photos **p40** Geoff Doré,
Bruce Coleman Ltd, (inset left) Geoff du Feu, Planet Earth
Pictures, (inset right) Jack Dermid, Bruce Coleman Ltd **p41**
Miami Tourist Board **p42** Walter Joseph, Bruce Coleman Ltd
p43 Hackenberg, ZEFA.

CONTENTS

INTRODUCTION

The first fires

The world's first fires began by chance when lightning and erupting volcanoes set dry vegetation alight. Perhaps one million years ago or more, our early ancestors discovered that fires could keep them warm at night and scare off wild beasts. The discovery of fire meant that Stone Age hunters could spread north into colder lands. People also gradually learned how to use fires to cook food, and to smelt metals.

Early people learned how to use fire to extract metal from ore. This Egyptian shield (right) is made of bronze. In this modern forge (below), fire is still used to heat metals.

More than 200 years ago, Europeans began to burn more concentrated fuels than wood. By the late 1800s, furnaces burned coal in order to fuel the steam-engines that drove factory machines and railway locomotives. In the 20th century, petroleum oil powers road vehicles, ships and planes. Coal, oil and natural gas are burned in power stations, releasing energy that produces electricity.

Our civilised way of life in the modern world depends upon burning fuels. Yet there is truth in the old saying: 'Fire is a good servant but a bad master'. Since prehistoric camp fires first set trees alight, fires have sometimes run amok. Over the centuries, fires started by people have burned down huge tracts of forest and many buildings, killing millions of animals and people. Each year the USA alone suffers more than three million reported fires and about 10,000 deaths caused by fire.

One of the forest fires that devastated much of the Yellowstone National Park, Wyoming, USA, during the drought of 1988.

Coping with monsoon flooding in Calcutta, India

Water: friend or foe?

Like fire, water can be a friend or an enemy. Everyone needs water to drink. Farmers must water their animals and irrigate their crops. Factories consume water in huge quantities. No wonder people have built so many villages, towns and cities near rivers or on coasts.

Sometimes these settlements are threatened by the very waters they depend upon. This happens when water rises above its normal level. Heavy rain can turn mountain streams into raging torrents. Rain-swollen tributary streams can fill broad rivers until they flood their plains. Storm winds can drive seas across low-lying shores. Sea and river floods have wrecked towns and farmlands, drowning animals and people in their thousands.

The next chapters describe some truly devastating floods and fires. But this book also shows how people have repaired the damage, and how they have often found ways to limit the devastation or to prevent a similar disaster from happening again.

CHAPTER ONE

FLAMES IN THE FOREST

A major forest fire is one of the most terrifying and destructive of all natural disasters. Forest fires kill millions of plants and animals, burn down buildings and leave soil exposed to the damaging effects of wind and rain. Some of the worst forest fires in the world affect parts of North America, southern Europe, and Australia.

The Peshtigo fire

'The atmosphere was all afire.'

'It came in great sheeted flames from heaven.'

This is how two survivors described the Peshtigo fire, one of the most destructive forest fires ever to ravage part of the USA, in 1871.

After a particularly dry summer, small fires kept breaking out in the northeast of the state of Wisconsin. Late on 8 October, a fierce wind blew up and drove a wall of flame across that corner of the state. The people of the town of Peshtigo heard a frightful roaring sound and watched as the sky turned red. Then sheets of flame raced in, high above the tree-tops, fell upon their town, and wiped it out.

The Peshtigo fire is only one of many that have swept through coniferous forest and dry scrub in North America. In 1825, another immense conflagration destroyed part of the Canadian province of New Brunswick. In the

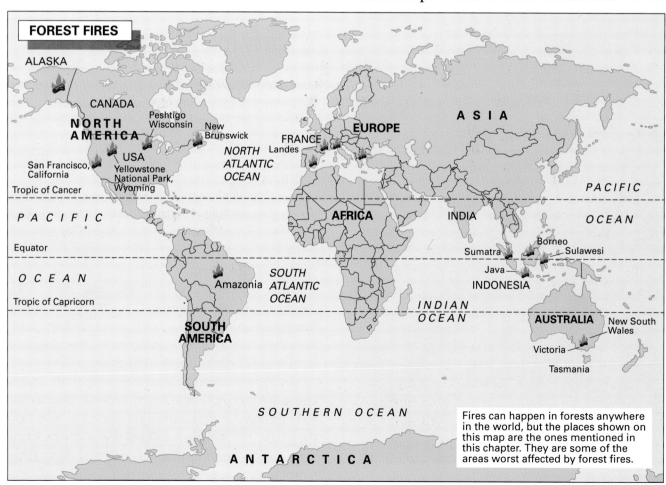

Fires can happen in forests anywhere in the world, but the places shown on this map are the ones mentioned in this chapter. They are some of the areas worst affected by forest fires.

Fire sweeps through dry undergrowth in southern France.

dry summer of 1988, fires blackened huge tracts of forest in Alaska, and in Wyoming's famous Yellowstone National Park. In October 1991, the worst fire since the earthquake of 1906 hit San Francisco in California, engulfing over 400 houses.

Fire in France

In hot, dry weather, dead wood and the resin-rich needle leaves of conifers form the perfect fuel, just waiting to be set alight. France, like North America, has large areas of coniferous forest which are liable to burn. Southeast and southwest France are the regions most at risk.

In the southwest, an immense sea of trees covers the area called the Landes, between the Bay of Biscay and the River Garonne. These conifers provide a useful crop of timber, turpentine and fuel. However, in August 1949, after a year of drought the Landes forests were tinder dry. When fires broke out south of Bordeaux, flames swept quickly west and north, surrounding villages and destroying 1300 square kilometres of trees. Eighty-three people were killed in this blaze.

Australian bush fires

Conifers are not the only trees at risk from fire. The leaves of some broad-leaved trees contain waxy or oily substances that burn furiously when set alight. Such trees include the eucalyptuses that grow on open forest land called 'bush' in Australia. Bush fires frequently break out in southeast Australia after hot, dry weather. Strong winds and the flammable eucalyptus oils often make these fires particularly deadly. While a fiery sea rages through the underbrush on the ground, great tongues of flame jump from tree-top to tree-top faster than anyone can drive a car. Sometimes a ball of flame takes off and soars above the trees to land and set fire to untouched forest farther on. People tell of fireballs leaping across the sea and burning islands two kilometres off shore.

When strong winds drive the flames, bush fires spread fast and far. In February 1851, the whole of the state of Western Victoria seemed to be ablaze. In 1939, 71 people perished in a single day in Victoria. In 1951 and 1952, bush fires blackened more than 24,000 square kilometres of the state of New South Wales. In 1962, fire killed 62 people in the south of the island of Tasmania. Late in 1990, the worst bush fires for 20 years threatened a suburb of the country's largest city, Sydney.

Preventing forest fires

Forest fires destroy more than a million hectares of trees every year in the USA alone. Worldwide, the amount of timber lost through fires is vast. It costs only a fraction of the forests' potential value to prevent these fires, or at least to put them out before they spread. Fire prevention can also help to save vulnerable houses built in areas of scrub or woodland that are liable to burn.

Preventing forest fires would be extremely difficult if they were all started by lightning. In fact, nine out of ten forest fires in the USA are caused by people who drop lighted cigarettes or matches on dry undergrowth. Human carelessness also sparks off most of the forest fires that break out in other countries. To try to prevent fires started by people, forest rangers

Clearing a firebreak to try to prevent a forest fire from spreading. Firebreaks can be used as tracks to drive deep into a forest (inset).

put up notices warning people not to light fires in the forest.

Another precaution is to cut 'firebreaks', or 'fire trails' through a forest. These are broad strips of ground that are cleared of trees and undergrowth. With nothing for the flames to feed on, the fire cannot spread from one side of the firebreak to the other. Some forests are criss-crossed by firebreaks that divide trees into many blocks of woodland. Sometimes strong winds may blow flames across firebreaks. But if firebreaks do not always stop fires spreading, at least they serve as tracks along which fire-fighting vehicles can drive deep into a forest to attack a blaze. Similarly, in the USA, Californian law says that undergrowth must be cleared away within ten metres of all buildings, and within 20 metres where the fire risk is particu-larly high. Experts even recommend installing high-pressure water sprinklers around build-ings. These can be used to prevent the sur-rounding vegetation from becoming dry enough to burn.

Fighting forest fires

Fighting forest fires has special difficulties. One problem is to spot a fire before it spreads. Another is to reach the fire quickly and with the right equipment. Forests often sprawl across remote hillsides where there are no lakes or rivers to supply water for dousing the flames.

Early warning of a fire can be given in several ways. Forest rangers scan huge tracts of forest from tall towers. Spotter planes or helicopters may fly over a forest every day, and satellites orbiting around the Earth can show where forest fires have broken out. Once alerted to a forest fire, local fire departments send out trucks capable of moving over rough, unsur-faced tracks. These fire trucks hold their own supplies of water and chemical foam to help subdue a blaze. The fire fighters' first aim must be to stop a fire from spreading. If possible, they begin by spraying water or chemicals upon the advancing flames to slow them down.

If the fire is burning far from any track, planes or helicopters may drop water bombs, or spray the fire from the air in order to damp it down. Sometimes fire fighters drop by parachute to dig a firebreak ahead of the blaze.

Making firebreaks can play a vital part in killing off a forest blaze. Besides spreading from tree to tree, forest fires feed on the dead leaves, twigs, branches and logs cluttering the forest floor. Fire fighters use axes, shovels and even bulldozers to clear away a broad belt of this fuel, and to scrape away the topsoil. They also fell trees standing in this belt. Then they may start fires between their firebreak and the forest fire itself. This controlled burning of undergrowth and litter broadens the firebreak until it grows too wide for the fire to jump across.

As the fire runs out of fuel and dies down, fire fighters may smother it with soil to kill it off. Then they remove all flammable material from the edges of the burned area, so that the hot embers cannot set the land beyond alight.

Aeroplanes and helicopters are used to control forest fires.

Straw and stubble

After harvest, many farmers burn off the straw and stubble left behind in the fields. In order to stop the fire spreading out of control, strict rules are laid down about the size of the area being burned, and the strength and direction of the wind.

Many people object to stubble burning because of the danger to wildlife, the nuisance of the smoke and ash, and the waste of a natural resource – straw – which can be used for other purposes. Straw can be treated with alkali or ammonia to make it more nutritious for animal feed, or used as bedding for livestock. Straw-burning furnaces and boilers can provide heat for hot water and central heating. Straw can also be used in board and paper manufacture. In the EC, straw burning is already banned in some countries.

Saving by burning

In 1891, a group of Californians claimed that they had saved a grove of giant sequoias (the world's largest trees) from burning 29 times in five years. Gifford Pinchot, the director of the US Forest Service, wondered 'Who then saved them during the other three or four thousand years of their life?' Foresters came to realise that fairly frequent forest fires actually help sequoias and some other kinds of trees to grow. Their thick bark can withstand the flames, and the heat opens the trees' cones, allowing the seeds to drop out. Forest rangers learned something else as well. In areas where fire killed off black spruce forest in Alaska, birches and other deciduous trees sprang up in its place, providing food for moose and other animals. So Alaska's forest fires increased the variety of trees and animals.

By the 1960s, such discoveries persuaded American forest managers to allow forest fires sparked off by lightning to burn unless they threatened human life or property. Indeed, foresters began deliberately burning areas of forest to stop thick layers of litter fuel collecting. Controlled burning became unpopular and was stopped after the savage North American forest fires of 1988. But the reasons for it are still sound.

Elsewhere in the world, fires have altered or destroyed some forests for good. Many of these fires were deliberately lit by people. Australia's fire-resistant grasslands and eucalyptus forests have replaced old vegetation that was burned off thousands of years ago by Aborigines. Africa's savanna grasslands were largely forested until prehistoric hunters burned the trees down. In the same way, fire was used to replace trees with grass in parts of the North American prairies, and the steppes of Central Asia.

Now, man-made fires are helping to destroy the great rainforests of the tropics – places normally too moist to burn. In Indonesia, local farmers burn off the land ready for wet-season planting. In 1991, after four months of drought, the fires spread, setting alight huge areas of rainforest on the islands of Sumatra, Borneo, Java and Sulawesi. The smoke from the fires blotted out the sun over the region for days.

Amazonia ablaze

The largest area of rainforest at risk from fire is the vast tract of tall, broad-leaved evergreen trees covering the basin of the River Amazon in South America. This forest is vanishing for two main reasons: ranchers want the land for raising cattle, and poor peasants need land for growing crops. Loggers, miners and charcoal smelters have also staked their claims to the forest land.

In 1987, an area of Amazonian forest about the size of New York State, or Czechoslovakia, went up in flames. On one September day, the US weather satellite NOAA-9 spotted no fewer than 7600 different fires alight in Amazonia. At such times, local aircraft were grounded: their pilots could not see to land through the fog of smoke shrouding Amazonian South America, from the Atlantic Ocean to the Andes Mountains. The Brazilian environmental scientist, Alberto Setzer, calculated that 170,000 fires had

Giant sequoias can benefit from forest fires.

The native peoples of the Amazon rainforest (above) have learned to live in the forest without damaging its delicate ecosystem.

A photograph taken from the American Space shuttle (left) in September 1988 shows the whole of the Amazon Basin obscured by smoke from forest burning.

ravaged much of Amazonia that year alone. Setzer's use of satellites proved that the Amazonian forest was vanishing far faster than anyone had thought.

Saving the rainforest

This scientific proof of the damage done by burning Amazonia began to alarm people everywhere, especially when scientists showed that, without its trees, the forest soil soon lost its productivity. The croplands or grasslands that replaced the forest quickly gave way to useless scrub. A leading Brazilian environmentalist, José Lutzenberger, calculated that an area of rainforest yields ten times more food (fish, game, fruits and nuts) than the same area once it has been burned down and used for ranching. Even the Brazilian government began to see that burning the rainforest was not the best way to provide poor peasants with farm land. The government set up Operation Amazonia to preserve the forest against illegal acts by charcoal smelters, loggers, ranchers and gold prospectors.

Repairing the damage already done in Amazonia is difficult. Tall rainforest trees may never grow again on land where the forest has been destroyed by fire because the soil has lost its nourishment. The best hope must be to make the wisest use of the rainforest that still remains. This seems to be by encouraging the traditional ways of life of small local groups who gather forest crops. Among these are the *seringueros*, or rubber tappers, and the Amerindians who clear and cultivate small patches of land, and then move on to allow the soil to recover.

Many people fear that big business interests could still wreck Amazonia. Damage even comes from schemes to put back trees to replace the forest already burned down. For instance, the immense Jari River project replaced hundreds of kinds of slow-growing, native hardwoods with plantations of a few kinds of fast-growing conifers, eucalyptuses and other species. Every few years, these trees are cut down as crops. But Amazonia's rich variety of animals cannot live among these foreign species, and damaging diseases can spread fast through such plantations.

CHAPTER TWO

CITIES ON FIRE

Blazing buildings

Since camp fires first set alight prehistoric shelters, buildings have been catching fire. The bigger the buildings, the more terrible the blazes tended to be. For centuries, the largest and grandest buildings were cathedrals and churches, and the worst fires occurred when they caught alight. Indeed, many European churches have been partly burned down and repaired since early medieval times.

Perhaps the deadliest of all church fire disasters occurred in 1863, in a Jesuit church in Santiago, Chile. Some records say that 2500 people perished. On 8 December 1863, the church was brilliantly decorated for a great religious festival. As the last oil lamps were lit, one end of a wood-and-canvas model of a crescent moon caught fire. At once, flames darted through the paper garlands decorating the altar and reached the great muslin curtains hanging from the wooden roof.

Panic followed. The congregation surged towards the main door, the only way out. A few women tripped and fell. More toppled on to them from behind. People outside watched helplessly what a Santiago newspaper called: 'the most harrowing sight that ever seared human eyeballs'. Most of the congregation died in the fire.

Modern buildings

In more recent times, more modern kinds of buildings – factories, warehouses, night-clubs, hotels and offices – have suffered some of the worst single-building fire disasters. As in Santiago, highly flammable materials contributed to these disasters. In 1970, a French night-club near Grenoble was the scene of a catastrophe when crinkly plastic covering the inside walls of the big hangar-like building was set alight by a

Fire-safety laws have been tightened in discos and night-clubs since the fatal fire at the Club Cinq-Sept near Grenoble, in 1970.

burning match. The stuff began to burn and melt, giving off dense clouds of smoke. Moments later a huge flame swept through the club, and choking smoke billowed down on scores of dancing couples. Nearly 150 young people perished in this tragedy.

In 1974, many office workers lost their lives when fire gutted the 25-storey Joelma office building in São Paulo, Brazil. When the eleventh floor caught fire, flames fed upon the paint and plastics of floors and window frames. Helicopters rescued about 100 people from the roof, but more than 220 lost their lives.

Lessons taught by terrible disasters such as these have led to the invention of new fire-fighting methods (see page 16), and fire-safety laws (see page 17).

City conflagrations

In past times, fires frequently destroyed large parts of cities where tall wooden buildings huddled close together. One of the most famous of these city fires was the Great Fire of London in 1666. Thanks to a famous diary, this is the best-described of any city conflagration.

The Great Fire of London, in 1666. Old St Paul's stands in the centre of the picture. The progress of the fire was recorded in the diary of Samuel Pepys (inset). His famous diary covers the period between 1660 and 1669 and also includes a vivid account of the Great Plague.

The Great Fire of London

Early on 2 September 1666, fire broke out in a baker's shop in Pudding Lane, an alley running down to the River Thames. Flames quickly spread through timber houses nearby, which were bone dry after a long, hot summer. Several alleys of buildings were soon ablaze.

Next day, from a boat, a government official called Samuel Pepys watched people abandon their homes to the flames. Pepys saw:

'Everybody endeavouring to remove their goods, and flinging them into the river, or bringing them into lighters [barges] that lay off.'

Pepys delivered to the Lord Mayor of London the King's order to pull down houses, in order to stop the fire spreading. But Pepys's diary claims the faint-hearted mayor simply cried:

'Lord! What can I do? I am spent: people will not obey me. I have been pulling down houses, but the fire overtakes us faster than we can do it.'

Later that day, Pepys and his wife watched the fire from a boat:

'So near the fire as we could for smoke; and all over the Thames, with one's faces in the wind, you were almost burned with a shower of fire-drops . . . . When we could endure no more upon the water, we to a little alehouse on the Bankside [south bank] . . . and there staid till it was dark almost, and saw the fire grow; and as it grew darker, appeared more and more; and in corners and upon steeples, and between churches and houses, as far as we could see up the hill of the City, in a most horrid, malicious, bloody flame . . .'

Worse followed. Sparks flew from the main fire through the air and set distant buildings alight. On 4 September, fire reached the greatest building of all: old St Paul's Cathedral. The church roof caught alight and fell in. The fire blazed for four days and nights. When the flames died down, most of London lay in ruins. Eighty-five churches and more than 13,000 houses were destroyed. More than 100,000 people were homeless.

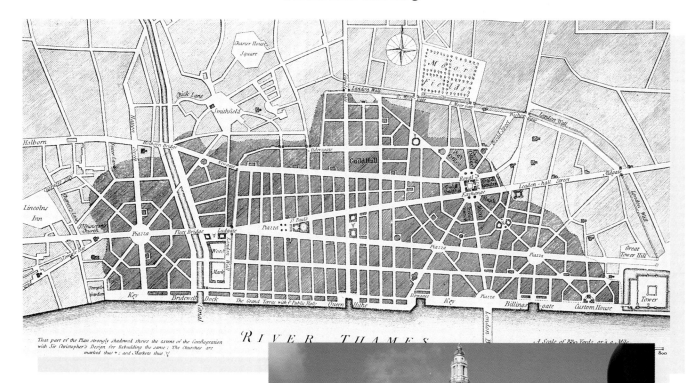

That part of the Plan strongly shadowed shews the extent of the Conflagration with Sir Christopher's Design for Rebuilding the same : The Churches are marked thus + : and Markets thus ⦂

Only five days after the end of the Great Fire of London, Sir Christopher Wren presented the King with his plan for rebuilding the city. Instead of the tangle of narrow, twisting streets which had been destroyed by the fire, he proposed an ordered city with stone buildings constructed along straight streets of three different widths.

Sir Christopher Wren designed more than 50 new churches for the rebuilding of London, among them St Paul's Cathedral as it stands today.

Rebuilding the city

While London's ashes were still warm Christopher Wren was one of four architects to offer the King a complete scheme for rebuilding the city. The King accepted none of them, but he made sure that no fire should sweep the whole city again. A royal proclamation declared that new buildings must be built of brick or stone, with fire-resistant walls between terraced houses. Main streets should be too wide for flames to cross, and, instead of houses, there should be a broad quay along the river so that fire fighters could easily take water from the Thames. Rebuilding took some 20 years and the new London went up piecemeal, not according to Wren's masterplan.

The Great Chicago Fire

North America's most famous conflagration was the Great Chicago Fire of 1871. In under 40 years, Chicago had grown from a tiny settlement to a great railroad centre with banks, theatres and blocks of offices. But, as in London two centuries before, many buildings were built of wood. In 1871, they were tinder dry after a hot and unusually dry summer.

Late on 8 October, fire broke out in the southwest of the city. Strong southwesterly winds soon drove the flames through wooden buildings, north and east across the city. Coal and timber yards, distilleries and grain stores all burned furiously. Six-storey buildings blazed and vanished in five minutes. Iron girders in a 'fireproof' bank became so hot that they bent until the walls and ceilings of the building bulged and collapsed.

Journalists reporting for the *Chicago Daily Tribune* told how:

'Flames would enter at the rears of buildings, and appear simultaneously at the fronts. For an instant the windows would redden, then great billows of fire would belch out, and meeting each other, shoot up into the air a vivid, quivering column of flame, and poising itself in awful majesty, hurl itself bodily several hundred feet and kindle new buildings . . . The whole air was filled with glowing cinders, looking like an illuminated snowstorm . . .'

The Great Chicago Fire raged for 27 hours until rain fell and put it out. At least 300 people died, more than 90,000 lost their homes, and 18,000 buildings were destroyed.

Chicago revived

The city's destruction gave the chance for a fresh start. By the 1880s, leading architects such as William le Baron Jenney and Louis Sullivan were designing new buildings. Among these was the world's first skyscraper, with an iron frame which bore the weight of the building. Thanks to the designers of the so-called Chicago School of Architecture, the rebuilt city rose from its ashes better planned and constructed than before.

A scene of devastation (left) along Chicago State Street in the aftermath of the fire in 1871, and Chicago rebuilt (above) – as it looks today.

One of the leather buckets used to fight the Great Fire of London in 1666.

Modern fire fighters make use of extending ladders to attack a blaze (right), and fire-resistant suits to protect themselves from heat and fumes (below).

Fire-fighting

People have come together to fight fires ever since towns were first built about 10,000 years ago. However, fire fighters were not organised and trained until much later. Old records hint at the activities of fire brigades in China and Egypt, more than 4000 years ago. But the earliest well-documented fire brigade appeared in Italy after fire had badly damaged Rome in AD 6. To save his capital from more fires, the Emperor Augustus organised 7000 full-time fire fighters, called *vigiles*. Some of the men acted as water-carriers, others directed hand-operated water pumps. Their tools included axes, brooms, buckets, ladders and poles. Fire-fighting tools and methods remained little changed until the 1500s, when the first fire engine on wheels was used in Augsburg, Germany.

Modern fire fighters are highly-trained experts who operate in teams. Each team usually deals with fires inside its own area, but a control centre may call up teams from several areas to help put out a large blaze. Fire fighters go to a fire equipped with fire-resistant clothing, and helmets, boots and gloves to guard against injury from flames and falling debris. Axes, bolt-cutters, crowbars, sledge-hammers and other tools enable them to break down doors to rescue people trapped inside a burning building. Breathing apparatus and aluminium-coated heat-reflective suits allow fire fighters to walk through flames and poisonous smoke.

Fire prevention

Putting out a blaze is a hazardous activity, even for highly-trained fire fighters. Preventing fires from starting, or ensuring that they do not spread quickly is, therefore, extremely important. In most countries there are laws and regulations to make sure that buildings, clothes and furnishings are as safe as possible. Some building materials and structures must be able to resist flames for many minutes without collapsing. Other regulations say that certain kinds of materials used for furnishing and clothing must not catch fire easily. Fire-retardant coatings can be chemically bonded to the fabrics of clothes, carpets, curtains and upholstery. These coatings help to slow the rate at which fires spread. Flames playing on this coating make it swell to form a layer of foam that insulates the substance beneath. There are also fire-retardant materials that give off gases which suffocate the flames.

Safety in the home

Most deaths from fire happen through carelessness in people's homes. Among the basic safety rules are these:

— Never leave young children alone near cookers or open fires, and never let them play with lighters or matches.

— Avoid overloading an electric socket with a number of appliances, and ask an electrician to replace any electric wiring that is worn.

— Throw out old clothes or other flammable items instead of storing them in a place where they might feed a fire.

— Keep paint, petrol or other flammable liquids in firmly closed containers in a shed or garage.

— Place clothes and soft furnishings well away from open fires or portable heaters.

— In the kitchen, never fill chip-pans more than one third full of cooking oil or fat, and keep paper towels away from cooker rings.

— At bedtime, switch off and unplug any electrical appliances not meant to be left on, safely dispose of burning cigarettes, place a guard around any open fire, and shut the doors of empty rooms.

— Smokers should never smoke in bed.

In one fire test, it took just three minutes 20 seconds for a small fire in one corner of a sofa to engulf a living room in flames and smoke.

30 seconds after ignition

2 minutes 15 seconds after ignition

3 minutes after ignition

3 minutes 20 seconds after ignition

Building materials

Building materials are divided into two categories for fire protection: combustible and non-combustible. Materials are tested by plunging a small sample into a furnace heated to 750°C. If they give off flammable gases, or burst into flames for ten seconds or more, or if the temperature of the furnace rises by more than 50°C, then they are called combustible. Combustible materials can sometimes be made more fire resistant if they are treated with fire-retardant materials, such as flame retardant salts or gypsum plaster. However, despite these treatments, combustible materials can never be made non-combustible because they will continue to provide fuel for a fire, and produce smoke and toxic gases.

Non-combustible materials do not allow flames to spread, and they do not provide fuel for the fire. But non-combustible materials are not necessarily good at resisting fire – glass transmits heat and shatters easily, allowing smoke and flames to spread, and steel expands and buckles in intense heat.

Some examples of combustible materials include timber, fibre board, cork, wood-cement chipboard and all plastics and rubbers. Non-combustible materials include asbestos cement, glass, brick, stone, concrete and metals.

When a building is being designed, adequate protection against fire must be incorporated into the plans. In a fire, some materials may melt, burn, crack, expand or lose strength; others may give off smoke and toxic fumes that are far more dangerous to people than the fire itself. Different parts of a building are protected in different ways, depending upon their function and the design itself.

Many buildings are constructed around a steel frame. If the steel is subjected to the intense heat of a fire then the frame will eventually collapse, so it is important that the frame is protected to give fire fighters enough time to put out the fire and to allow people to escape. A steel frame can be encased in fire-protection board, or covered in a fibre spray, to form a physical barrier between the fire and the steel. The frame can also be covered with a type of paint that reacts to heat by forming a charred solid, insulating the steel against the fire.

Flammability tests

Preventing a fire is much easier when you understand why a fire takes hold and spreads. Scientists in fire research establishments find out how different substances and structures burn. Experts from all over the world exchange their findings. This helps governments to plan their safety laws.

Among the chief fire-study centres is the Fire Research Station at Borehamwood near London, and its fire laboratory at Cardington in Bedfordshire. This is the world's largest fire laboratory: a colossal building that was once an airship hangar. Dwarfed inside stands a full-size three-bedroomed brick-built house, and rows of tall storage racks like those in warehouses. Furnished rooms in the house and boxes on the racks are deliberately set alight to find out how different substances and structures burn. Studies have revealed how dense and poisonous the smoke can be when modern furnishings catch fire (see page 17).

The fire researchers also devise computer-generated images showing how a fire spreads through a large building. After a major fire in a building such as a night-club or a football stadium, fire-research scientists often visit the site of the disaster in order to study how the fire began and spread.

Another example of the kind of safety test performed by fire-safety experts involves the fabrics used for nightwear. Many children and old people have died from burns when their nightclothes caught alight from coal or electric fires. In many countries there are now strict flammability standards, and manufacturers must label most kinds of nightwear to show if they reach these standards.

Fire studies like these help governments to set fire-safety standards for new kinds of clothing, furnishing and building materials. Fire tests made by the scientists can then show manufacturers if their products are safe enough to sell to the public.

The huge structure that houses the Fire Research Laboratory at Cardington in Bedfordshire, England, was built as an airship hangar. It now contains a house, rows of storage racks and a mock-up aeroplane, all of which can be set alight to find out how different structures and materials burn.

CHAPTER THREE

FIRE ABOVE AND BELOW

Fires underground, in aircraft or at sea are among the most terrifying fires of all. Some happen because people have not understood the risks involved. In other cases, fires break out when the risks are known. This can happen if there is no safety system, or if safety officials are inefficient, or even if there is too little money to spend on safety equipment. The worst disasters have forced planners and manufacturers to pay more attention to safety in tunnels, aircraft, ships and marine oil rigs.

Mines and tunnels

Thousands of people have died in fires in mines and tunnels. From time to time, explosive fires occur deep down in coal mines and gold mines. A mine fire usually starts because a spark sets fire to dust in the air, or to methane gas escaping from the rocks. In 1986, a welder's flame accidentally set off one of the world's worst gold-mine fires, in South Africa. About 180 miners died in the Kinross Gold Mine. In 1942, more than 1500 Chinese miners died in the world's worst-ever coal-dust explosion.

The miner's safety lamp in use in the 1870s

Safety precautions now make such major mine disasters rare. One simple but important early safety measure was the miner's safety lamp invented in 1815 by the British chemist, Sir Humphry Davy. Although the lamp burned oil, a fine wire gauze around its flame prevented the heat of the flame from igniting any methane gas in the air. A pale blue halo around the ordinary flame also gave warning of lethal methane gas. Such lamps are still in use. In modern mines, fans force or suck fresh air through ventilation ducts into every cranny of the shafts and tunnels in order to stop methane gas becoming concentrated. Sprinklers spray water to damp down dust that may be floating in the air.

Miners are the people most at risk from underground fires, but fires have also broken out in railway tunnels. In 1987, a fire started during the rush-hour in King's Cross underground station in London. A young woman standing on a rising escalator later said:

'When I was about half way up I looked down . . . There were jets of smoke coming out from gaps between the escalator steps. I could see a big orange glow underneath the escalator.'

She warned staff at the top, but five minutes later fire and smoke rushed up the escalator and a wall of flame swept through the ticket hall. One man who survived said:

'There was smoke everywhere, white smoke, then thick, black horrible smoke. It was pandemonium.'

Soon, 200 fire fighters were tackling the blaze, but they came too late for 31 travellers who were overwhelmed by fumes and flames.

Investigations later showed that the King's Cross fire had begun when someone dropped a burning match on to an old escalator with

Escalators at King's Cross underground station, London, after the fire in 1987.

wooden steps. The fire's 'flashover', or sudden spread, might have happened when the plastic lining of the ceiling caught fire. Replacing flammable plastic and wooden panelling was one of a long list of safety measures already drawn up by the head of London's underground system after an earlier fire at another station. Another problem was the amount of uncollected rubbish underneath the escalator, including fluff and dirt soaked in oil from the machinery. This helped the fire to burn fiercely.

The King's Cross fire taught many safety lessons. Here are some of them: stop people smoking underground. Replace wooden panels and skirting boards with metal. Prevent flammable rubbish piling up underground. Install smoke and heat alarms that automatically trigger sprinklers to put out a fire before it spreads.

A tunnel under the sea

By the early 1990s, fire brigades in southeast England and northwest France had something new to worry about: the Channel Tunnel. This huge engineering project involves three tunnels, 37.5 kilometres long, which lie side-by-side beneath the English Channel, connecting France and England. Trains will run through two of the tunnels, the third is a 'service tunnel' into which passengers would escape if there was an emergency. Together, the two running tunnels can hold up to 10,000 passengers at once, in several trains. Truck drivers will probably travel separately from their vehicles, but most people will stay sitting in their cars or

coaches which will be parked in the moving railway wagons. Cross-passages connect both running tunnels to the smaller service tunnel in-between.

The tunnel's planners know that a fire inside the tunnel could be as dangerous as any mining fire disaster. They have taken considerable care to reduce the fire risk. Despite the fact that in a similar tunnel – the Simplon Tunnel that runs beneath the Alps – there has been only one small fire in the last 40 years, the safety standards in the Channel Tunnel will be even higher than in the Alpine tunnels. If a fire breaks out inside a train, an early-warning system should detect the flames or smoke. Fire doors should slow down the spread of fire between one wagon and the next, long enough for the train to leave the tunnel. If necessary, passengers can escape into the service tunnel.

Several kinds of fire extinguishers would attack the blaze, including foam extinguishers which would smother a petrol fire. If these measures were not sufficient, small amounts of halon gas would be released into the wagons. Halon is very effective because it not only starves the fire of oxygen, it also stops the fire spreading by preventing the chemical reaction that forms the flames themselves.

Digging out one of the three tunnels that together will form the Channel Tunnel beneath the English Channel.

Danger in the air

Modern passenger aircraft hold two main kinds of loads: people, and the liquid fuel needed to power the aeroplanes' engines. When fully-loaded aircraft crash, or crash-land, spilt fuel is liable to catch alight with terrible results for those inside the aircraft. Airports keep fire trucks standing by to tackle such disasters.

However, fire trucks could not prevent one of aviation's most devastating accidents. In 1977, two jumbo jets collided at Los Rodeos airport on the island of Tenerife. Exploding fuel tanks killed almost 600 passengers and crew. This tragedy happened at an unusually foggy and busy mountain airport, with only one of its three radio frequencies working. Poor radio communications might have led one of the two pilots to misunderstand the instructions from the control tower. Poor visibility and over-crowded runways also contributed to the disaster. The safety lessons to be learned were plain.

Little can be done for passengers in an aircraft already swept by flames. But flames or smoke may be detected in an aircraft cabin minutes before the fire grows uncontrollable. Then, quick well-planned action can save many lives.

The cabin crew can tackle a small fire with fire-extinguishers. If an aeroplane crash-lands, the cabin crew will help all passengers to leave as fast as possible.

Some pilots and cabin crew train for fire emergencies in specially constructed crash simulators. Using smoke, and even flames, these imitate the inside of an aircraft which has crash-landed in a tilted position.

Protection from fumes

In 1985, a Boeing 737 caught fire on a runway at Britain's Manchester Airport. Nine people died in the flames, but smoke and poisonous fumes killed 46 more people before they could escape. One survivor claimed that one breath made his lungs feel 'solidified', and two breaths made him feel like 'falling down asleep'.

Smokehoods might have saved some passengers from the fumes. Smokehoods completely cover the head and are sealed in at the neck. They protect people's eyes, mouths and noses from poisonous and irritating gases. After the

An airport emergency crew in training. The fire fighters are spraying foam to extinguish burning jet fuel.

Manchester accident, Britain's Civil Aviation Authority (the CAA) decided not to recommend the use of smokehoods. The main problem was the time it took to put the smokehoods on. The CAA felt that in an aircraft fire the first aim ought to be to get passengers out before flames engulf the aeroplane. But the CAA did lay down new fire safety rules. These rules declared that new aircraft must have more fire-resistant linings for their walls and ceilings, and more fire-resistant seats.

In May 1991, researchers from British Petroleum demonstrated a new method of firefighting. This new system uses a mist of water to extinguish burning fuel. The water is forced through a special nozzle which creates spray with the correct size droplets to reduce the temperature of the fire, and to absorb smoke and gases. If the droplets are too small they turn into steam; if they are too large they splash the fuel around, causing the fire to spread. Tests have shown that aircraft interiors fitted with water sprays remain cool, allowing passengers to survive for long enough to escape. The system has the added advantage of using very little water, making it suitable for aircraft, and possibly for the trains that will run in the Channel Tunnel.

Fire at sea

When all ships had wooden hulls fires often broke out at sea. That risk grew less with the development of metal hulls, modern fire precautions and fire-extinguishers. Fireboats, squirting powerful jets of foam or water, can now help to put out fires on ships in ports, and well away from land. A powerful fireboat pumps out more than 40,000 litres of water a minute.

Many lives could easily be lost in fires on oil rigs – those man-made metal islands perched above the sea on metal legs embedded on the ocean floor. Scores of people live and work on each platform, and dozens of platforms are dotted around the world.

The fire precautions on an oil rig ought to make a major oil-rig blaze impossible. If gas escapes, gas sensors should sound alarms. If fire breaks out, other sensors should switch on fire alarms and sprinklers. Fire walls give living-quarters some protection. The platform contains breathing apparatus, life jackets, survival suits and lifeboats as well as foam, halon gas and water to fight whatever kind of fire might start.

Fireboats are used to fight fires in port and out at sea.

Oil rig alert

Despite strict safety measures, fire and explosions killed 167 people when the Piper Alpha platform blew up in a British sector of the North Sea, on the evening of 6 July 1988.

The first hint of danger was the sound of escaping gas 'screaming like a banshee'. The leaking gas caught fire. Then came a series of explosions. Smoke and fire quickly engulfed the platform, and flames leapt 200 metres high. The heat grew so intense that the steel of the platform began to melt. Most men had been sleeping, eating or resting in Piper Alpha's four-storey accommodation block. One survivor later called it 'a hotel sitting on a potential fire bomb', for it stood above the drilling area where huge quantities of oil and gas were pumped up.

Workers obeyed safety rules telling them to stay inside the accommodation block until help came. But no rescue helicopter could land upon the blazing, buckled platform, and with the control room wrecked there was no way to send instructions to the trapped men.

Disobeying company instructions, other workers took their chance and jumped into the sea. One survivor said:

'It was fry and die, or jump and try.'

Another later told a newspaper reporter:

'The steelwork I was climbing down was red hot. I was dancing across. My hands were burned. I was in the water for about 20 minutes before being picked up.'

A long inquiry followed Piper Alpha's destruction. Investigators found that many mistakes had led to the disaster. A valve had been removed for repair but not replaced, allowing gas to leak out from a pipe. Sprinkler switches had not been set to come on automatically, and many sprinkler heads were blocked. When fire broke out, the design of the rig made it impossible for men to escape by helicopter or lifeboat.

The Piper Alpha oil rig after the explosion and fire in July, 1988

One of the 800 oil well fires in Kuwait that raged after the end of the Gulf War. Huge lakes of oil surround the well.

To prevent more fires like Piper Alpha, investigators insisted that each offshore oil-rig worker must have his own survival suit, life jacket, smokehood, fireproof gloves and torch. All workers must be trained for an emergency. Offshore oil rigs should include a fireproof refuge where men could shelter until rescued, or until the fire was put out. There should be escape routes to lifeboats. All companies must establish and maintain high safety standards which should be checked regularly by government inspectors.

These British improvements went some way towards reaching Norway's well-established safety levels. Norway's offshore oil platforms keep living quarters well away from the areas where oil or gas are piped aboard. No fewer than seven walls protect the living accommodation from production areas. Lifeboats kept below the living quarters can slip into the sea in seconds. If all oil companies maintained such high safety standards, the risk of future disasters such as Piper Alpha would be extremely small.

Burning wells

One of the biggest fires in history raged for over eight months in the oil fields of Kuwait. After the end of the Gulf War in March 1991, it was discovered that 800 out of the 950 oil wells in Kuwait had been set alight by the Iraqi invaders. Six million barrels of oil – almost ten per cent of the entire world's daily oil consumption – were going up in smoke every day, polluting the atmosphere for hundreds of kilometres and coating the surrounding regions in a 'black rain' of soot, sulphuric acid and other toxic substances. Early estimates by the famous Texan fire fighter, Red Adair, suggested that it might take up to two years to cap all the burning wells. However, the 27 teams of fire fighters from all over the world completed the job in eight months, capping the last well in November 1991. The work to repair the wells will take longer, and it is feared that many will have to be abandoned.

CHAPTER FOUR

FLOODING RIVERS

When the water vapour in the air condenses it may fall to earth as rain or snow. On the ground, the water may evaporate once again into the atmosphere, or it may soak down into the earth until it finds an outlet. Over hundreds of years, streams and rivers carve their valleys out of the land as they flow towards the sea. Sometimes, after a big fall of snow in the mountains, or after heavy rain, too much water is channelled into a river. Then it overflows its banks. From the tiniest brook to the grandest river, every stream on Earth is occasionally subject to flooding.

Violent and sudden river floods, called flash floods, sometimes sweep down through narrow mountain valleys. This happens particularly during or after heavy rainfall when the earth becomes waterlogged and unable to absorb any more moisture, leaving the excess water to turn small streams into raging torrents. In 1976, a torrential mountain downpour sent a wall of water roaring down Big Thompson valley in Colorado, USA. One hundred and thirty-nine holiday-makers perished. Each year, about 200 people die in flash floods in the USA.

To warn of such disasters, the American weather service studies thunderstorms shown by radar and weather satellites. The weather service issues a flash flood 'watch' if a big thunderstorm seems likely to get stuck in one position, pouring heavy rain upon the hills. A flash flood 'warning' follows when flooding has started, or is about to start.

On low-lying river plains, floods rise more slowly but drown far larger areas of country-side. This flooding is the kind that causes most damage, for much of the world's best farmland and many of its great cities stand on low-lying river plains. Each year, at least one major flood disaster happens somewhere in the world.

Aid agencies work fast to save and help flood victims, but repeated river floods have forced people to find new ways of living with their wayward waterways.

The floods of Florence

Florence, in northwest Italy, lies on the River Arno. It is a city famous for its fine works of Renaissance art and architecture. On 4 November 1966, the River Arno burst its banks, and Florence suffered its worst floods for 600 years. Thousands of valuable ancient books and paintings were badly damaged or destroyed.

The trouble started when the northwest of Italy received one third of a year's usual rainfall in a single day and night. Worst hit was the Arno valley. By the morning of 4 November, the river was rising by one metre every hour. Soon it burst its banks, and a great surge of water up to five metres high swept through the city's narrow streets at up to 60 kilometres an hour. One eye-witness described the flood water as:

'A snarling brown torrent of terrific velocity, spiralling in whirlpools and countercurrents that send waves backwards. Down the first street back from the river, water is pent wall to wall within the canyon of the buildings. It plunges into the square in a vortex of waves, whirlpools and debris – branches, twigs, shoes, pocketbooks and paper, which swing around in a crazy bobbing dance.'

The torrent threw cars around like toys, smashed into thousands of small shops, and swamped 5000 houses. Flooding cut off the city's electricity, and ruined the water supply and sewerage systems.

Next day the water fell and people could take stock of the damage. The flood had wrecked 6000 shops and 10,000 cars, vans and lorries. The flood also left the streets of Florence deep in 500,000 tonnes of mud, mixed with sewage and the oil from smashed oil storage tanks. This mess swamped the ground floors of art galleries,

churches, libraries and museums. More than 1000 paintings, 300,000 books and 700,000 valuable documents were soaked or wore a coat of black, smelly slime.

People in countries everywhere were horrified. Hundreds of students from around the world flocked in to rescue rare books and manuscripts from the oily mud. These treasures went to different parts of Italy to dry in brick kilns and tobacco-curing sheds. Slow drying helped to save many flood-soaked paintings but it took many years to restore the damage.

The flood's destruction spurred Florentines to update their city. Workers modernised old buildings and put in a new sewerage system. At the same time, experts set about making Florence more flood-proof by rebuilding the river's artificial embankments. River authorities had worsened the flood by releasing water from dams in hills high above the city. This made it plain that city and river authorities must work together more closely.

The people of Florence and the rest of the world hope that such remedies will protect this famous city from its river in the future.

Reinforcing the banks of the River Arno (above).
People from all over the world came to Florence to help rescue and clean the many priceless works of art (below).

Mississippi floods

The Mississippi is North America's greatest river. With the Ohio, Missouri and other tributaries, the Mississippi drains more than one third of the land of the USA as it flows almost 4000 kilometres south to empty into the Gulf of Mexico. Halfway along, the river is more than one kilometre across where the Ohio flows in from the east. From here southwards, the Mississippi snakes in huge loops across a low, wide plain. As the river moves, it drops a heavy load of mud, which piles up on the river bed and on to the banks each time the river spills over. In this way, the mud builds up the river bed and banks high above the flat plain around the river. When the River Mississippi overflows, water spreads far across its floodplain, and the damage is sometimes immense.

In the 1820s, the naturalist, John James Audubon, saw floods around the Mississippi measuring up to 50 kilometres across. He wrote:

'The river rises until its banks are flooded and the levees [artificial banks] overflown. It then sweeps inland, over swamps, prairie, and forest, until the country is a turbid ocean, checkered by masses and strips of the forest, through which the flood rolls lazily down cypress-shadowed glades. . . .'

Mississippi floods did little harm before the European settlers arrived. The local native peoples knew how to live with the mighty river. They built their houses on high land, or on artificial hills, called 'Indian mounds' by the Europeans. However, as the European settlers established farms and cities on the river plain in the nineteenth and early twentieth centuries, the damage became worse. Each year, spring is the season of greatest danger. At this time melting snow and heavy rain add huge quantities of water to the upper reaches of the river and its tributaries.

Several River Mississippi floods are notable. One of the worst began in autumn 1926 and lasted to spring 1927. Floodwaters swiftly drowned an area of the Lower Mississippi Valley bigger than Denmark. The water grew so deep

The Gateway Arch in St Louis, Missouri, USA, with the Mississippi River in the background. The city lies just downstream from the point where the Missouri River joins the Mississippi.

that rescue boats were sailing at the same height as the tree-tops. Three hundred people died, huge areas of crops rotted, and roads and bridges were out of use for weeks. In 1937 came another massive flood. The river rose 17 metres at Cairo, Illinois, and the Lower Mississippi Valley briefly became the world's third largest lake. Only 135 people died, but a million others were forced to abandon their houses.

More big floods have happened since. In 1973, raging waters devastated land along 2200 kilometres of the Mississippi and its tributaries.

Taming the Mississippi

Ever since 1718, people have tried to stop the River Mississippi overflowing. That year workers built the first artificial bank, or 'levee', at New Orleans. Levees are wide embankments up to ten metres high, made of earth and sandbags, and sometimes reinforced by concrete blocks.

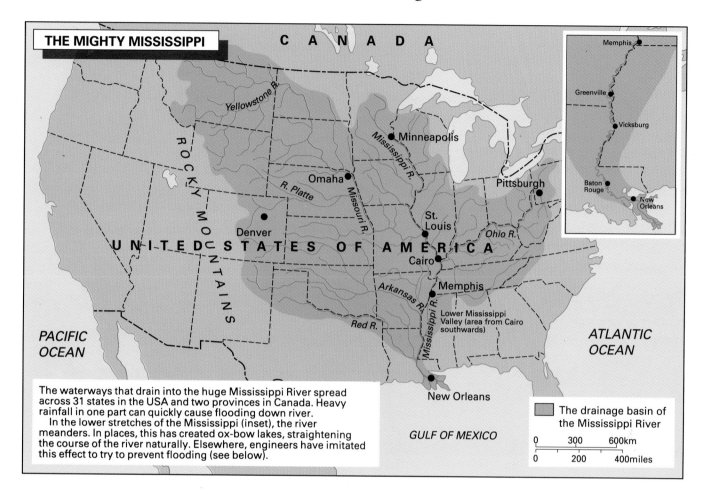

THE MIGHTY MISSISSIPPI

The waterways that drain into the huge Mississippi River spread across 31 states in the USA and two provinces in Canada. Heavy rainfall in one part can quickly cause flooding down river.
In the lower stretches of the Mississippi (inset), the river meanders. In places, this has created ox-bow lakes, straightening the course of the river naturally. Elsewhere, engineers have imitated this effect to try to prevent flooding (see below).

The drainage basin of the Mississippi River

| 0 | 300 | 600km |
| 0 | 200 | 400miles |

By the 1980s, 3200 kilometres of the Mississippi were lined by levees, and there were plans to add 4000 kilometres more. Levees help the lower river to hold extra water, so that it is less likely to overflow on to its plain.

Engineers have also come up with other ways to lessen flooding in the Lower Mississippi Valley. In its natural course, the river often loops back on itself with only a few metres of land between the loops. By cutting channels across the narrowest point between these meanders the course of the river is straightened, and the speed at which the water can flow increased. This means that floodwater is less likely to run over the banks. In places, engineers have also made spillways that carry overflowing water into special reservoirs. When the water level of the river falls again, powerful pumps push water from these reservoirs back up into the Mississippi. Other schemes slow down or hold back floodwater in the river's tributaries. People have built flood storage dams and planted trees on the slopes above. The trees prevent rainwater from running quickly into the river; instead the water soaks into the ground.

The US Government has spent huge amounts of money trying to tame the River Mississippi. However, some experts think that all these flood controls only put off until some future time a flood more terrible than any in the past. One reason is that the higher the levees are built the higher the river bed becomes, and the greater the disaster would be should the river overflow. Engineers have to keep on dredging mud from the river bed in order to stop it rising ever higher above the plain.

Some experts would rather solve the problem in another way. They argue that floods will happen whatever people do to stop them. Instead of trying to contain the river, they suggest that the US Government should let it overflow but make sure that the land at risk is used only for farming or sports fields that can survive floods now and then. Buildings already on this land could be moved, or the US Government could help to pay their owners to insure their property against the risk of flooding. No new towns or factories would be built upon the land at risk. Even so, established river towns would still need flood protection.

After two months of monsoon rains, flood water surrounds a village in Bangladesh.

Bangladesh at risk

Tucked into the northeastern side of India lies one of the world's poorest and most densely peopled nations, Bangladesh. No other nation is more liable to savage floods. Much of Bangladesh is a low delta which extends out into the Bay of Bengal. The delta is formed from the mud dropped by three great Asian rivers: the Brahmaputra, Ganges, and Meghna. More water flows through this delta to the sea than through any other river mouth except that of the River Amazon. Each summer, the three rivers brim with water from the heavy monsoon rains that fall on the Himalaya Mountains and the northern plains of India. The amount of water flowing into the rivers has increased as people have cut down the forests on the lower slopes of the Himalayas. Rainwater runs quickly down the bare mountain slopes, instead of being soaked up by the roots of the trees. In 1988, three fifths of Bangladesh was inundated. That year at least 2000 people drowned, and for the first time water swamped much of the capital, Dhaka.

Possible solutions

Bangladesh could not afford to repair the damage on its own, or to protect itself from future flood disasters. Food, medicines and other emergency supplies were rushed in by aid agencies from all over the world. Then, experts from many countries came up with ideas for long-term flood protection schemes.

French engineers suggested building 4000 kilometres of high banks to stop the flooding. But the team leader of the United Nations Development Programme thought that the floods should not be completely stopped. He claimed that controlled floods are vital for the rice and jute crops, and for inland fisheries which raise young fish and shrimps in flooded fields. Floodwater also keeps the soil moist and allows farmers to grow crops in the dry winter months. The United Nations felt that the best use for costly new embankments would be to protect people and places most at risk, especially the main cities, and key buildings such as hospitals and power stations.

Floods in Bangladesh: (top) a boy fishes for shrimps as flood waters recede, and (above) bundles of jute, a plant fibre used to make rope and sacks, are carried through flooded streets.

Blocks of concrete are used to reinforce the embankment of a river in Bangladesh.

Protecting Bangladesh

In 1991, a huge project to protect Bangladesh from the effects of floods and cyclones was begun. Called the Flood Action Plan, it is funded by the World Bank and is expected to cost more than nine billion pounds. Most of this money will be spent on improving river embankments inland in order to protect the cities and rural areas around the Ganges and Brahmaputra rivers from the worst effects of flooding.

Many people think that the money would be better spent on simpler measures instead of huge construction projects. These would include earlier warning of floods, the construction of flood shelters on areas of high ground, and adequate supplies of food and medical aid. Small-scale projects, such as houses built above the flood level on man-made or natural banks, are very effective. In some areas, special varieties of rice which can grow in high water levels are being cultivated, providing a source of food even if the land is badly flooded.

Using these simpler methods to combat the problem of flooding rivers in Bangladesh would mean that more money could be spent on flood defences in the coastal area of the delta, where cyclones and huge tidal waves regularly threaten the poor communities that live there (see page 38).

CHAPTER FIVE

DAM BURSTS

One of the most effective ways of storing a huge quantity of water is to build a dam. A giant wall is constructed across a river valley, and as the river water piles up behind the dam, an artificial lake, or reservoir, is formed. When the dam is full, outlets allow enough water to escape to keep the water level of the reservoir constant. Thousands of dams store water safely. But sometimes a reservoir bursts or overflows its dam, causing devastation as the water roars downhill.

When a dam fails, the resulting flood often swamps the valleys below. The failure of the Malpasset Dam in 1959 drowned 400 citizens of Fréjus, in southern France. More than 2200 people drowned in 1889 when a dam burst near Johnstown, Pennsylvania, USA. In 1976, some 30,000 people lost their homes in Idaho, USA, when the Teton Dam broke.

The worst floods caused by bursting dams have occurred in Asia. When the Manchu River Dam gave way in Gujarat, India, in 1979, about 5000 victims drowned. Some people think that as many as 900,000 people perished in 1938 when troops blew up a dam on the River Yangtze during a war between China and Japan.

Why do dams fail?

Such dam bursts have taught cruel lessons. Most of the dams that have failed did so because their designers had not taken enough care about where and how they were built. It is now known that the Teton Dam was filled with earth and rubble in such a way that water seeped right through and undermined it. The Malpasset Dam collapsed as its foundations crumbled. Several dams have failed because they were built on unstable ground such as broken rocks, or a fault or crack in the Earth's crust. If the land on one side of a fault should move, a dam built across that fault is likely to break.

Often, repairing a burst dam is not a good idea because the dam should not have been built in that position in the first place. At least past disasters can help by showing the builders of new dams the mistakes they should avoid. Nowadays, all countries have strict rules to govern how big dams are built. To this end, engineers work closely with geologists. These scientists make detailed studies of the rocks on which the dam will stand. They examine geological maps showing the rocks beneath the soil and they drill holes to collect rock samples from deep below the ground.

Engineers plan the construction of a dam according to the type of situation. A thin, curved concrete dam is usually best for blocking a gorge with a floor and walls of strong, solid rock. The bulging side of the curve faces upstream in order to withstand the huge pressure of the water in the reservoir. However, a slim concrete dam would collapse if built on the soft, weak rock called shale. On such rock, engineers build a low but immensely thick dam filled with earth and rock.

Engineers also study weather records to learn about the pattern of rainfall in the region. They can then design spillways large enough to carry off excess water from the heaviest rainfall likely in that area. This should prevent water from ever pouring over the top of the dam.

In spite of these precautions, accidents still happen. Some experts think that there ought to be a flood map for every dam to show the area of land likely to be inundated if the dam failed or overflowed. Computer programs can help cartographers to produce such maps. Experts also think that there should be a plan for warning and evacuating everyone at risk from floods. Special instruments can automatically record landslides into reservoirs, early signs of dam collapse, or rainfall heavy enough to cause a reservoir to overflow its dam. When any of these dangers threatens, automatic alarms could

Dams made of earth or rubble are used in broad river valleys. These are the only dams that are safe if the river bed is unstable.

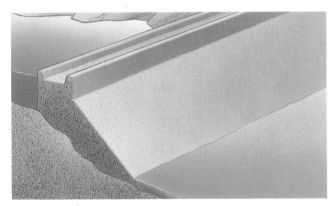

Gravity dams are built in narrow valleys, and use their huge weight to resist the pressure of the water.

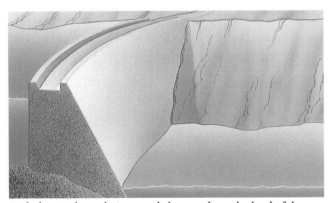

Arch dams rely on their curved shape to bear the load of the water.

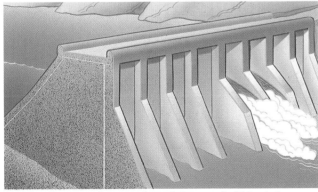

Buttress dams use their weight and shape to hold back the water. They can also withstand small earth movements.

alert the water authority. Then mobile loud-speakers and sirens could warn the people most at risk. Studies show that, for a town of 10,000 people, a good warning system can reduce the loss of life from 250 to two people.

The Canadian province of British Columbia already has inundation maps and plans. By the late 1980s, Australia had an inundation map for the Googong Dam: if that overflowed, flood water would cover parts of the Australian capital, Canberra. However, most countries still have few, or no, flood maps and plans.

The Vaiont Dam disaster

In one of the world's worst dam disasters the dam itself did not give way at all. This tragedy struck a mountain valley in northern Italy in 1963. The Piave Valley held a long, deep artificial lake, called the Vaiont Reservoir. The Vaiont Dam, completed in 1960, was the second highest in the world. Its curved concrete wall was about 260 metres high, and 190 metres across at the top. The dam stood on strong, hard, solid limestone rocks. But the slopes on each side of the reservoir were formed of layers of weak limestone and soft clay which sloped into the water. As the level of the reservoir slowly rose, the water soaked the clay layers of the slopes and made them slippery. A great mass of clay and limestone began creeping down the mountainside towards the lake. For three years the mass slid so slowly that no one noticed how much land was on the move. But, by September 1963, the slope was sliding downhill at the rate of 25 centimetres a day. Grazing animals, sensitive to the motion of the ground beneath their feet, left their pastures.

Suddenly, on 9 October, a huge slab of the mountainside was torn away and crashed into the lake. Some 240 million cubic metres of rock and soil thundered down into the reservoir, driving a mass of water 260 metres up the opposite slope. Then a huge wall of water 100 metres high surged across the top of the dam and hurtled down the Piave Valley. This mighty wave flattened five villages, sweeping before it everything in its path. One lucky survivor described what happened in these words:

The Vaiont Dam and, beyond, Mount Toc

The scene of devastation downstream

'I heard a terrible noise, something between thunder and a salvo from an artillery battery. I thought it must be Toc [the mountain] – the lights on the mountainside went out and a terrible wind began to blow across the lake and then came the wave. Then a thick dusty fog rose everywhere. It was dark and you couldn't see a thing.'

Terrible destruction raged downstream for many kilometres. In just seven minutes, 2600 people drowned. Yet, amazingly, the dam itself remained almost intact.

Like most dam bursts, this disaster should have been prevented. Old scars, showing where slabs of mountainside had come away, already hinted at the dangers of a landslide. Careful study of the unstable sloping rock layers should have shown what would happen once the water from the reservoir had soaked the rocks and made them liable to slip and slide. The Vaiont Dam disaster taught a lesson that builders of such dams will not forget.

An ice dam formed by a glacier in northern Norway. In summer, melting ice swells the lake, which drains into a river.

Super-dams

When a dam bursts, the land around is flooded accidentally. But when large dams are built the land behind the dam is flooded by the water being stored. In some countries, huge areas of farmland have been drowned by new reservoirs, and local people have had to leave their homes.

Since the 1950s, more than 250 'super-dams' have been built across the world. These huge dams are built to exploit the hydro-electric potential of the world's big rivers and deep valleys. The price of building these dams is high. In the Amazon Basin in Brazil, a programme of dam construction, including 12 super-dams, is expected to flood more than 26,000 square kilometres, destroying the rainforest environment and forcing many thousands of local people to move.

Ice dams

About two million years ago, a mass of ice many hundreds of metres thick spread south through North America. An ice tongue poking from this ice sheet blocked the Clark Fork River in what is now western Montana. River water rose behind the dam of ice. In time the trapped water filled a long, deep valley to form a narrow prehistoric lake, now known as Lake Missoula.

Periodically the climate warmed up, and the ice sheet shrank. This eventually caused the lake to burst through its ice dam. There followed one of the greatest floods the world has ever known. Up to 40 cubic kilometres of water an hour roared across the Columbia Plateau in what is now the western state of Washington. The flood water scoured a maze of deep channels in the flat rock, and dumped silt and gravel that show up from the air as giant ripples. These features of the so-called Channeled Scablands are found nowhere else on Earth.

Now, as then, ice dams in places as far apart as Alaska and Argentina could cause damage if they were to burst. In the Andes, naval planes have bombed a glacier to allow the lake water to escape. In Norway, engineers have made tunnels in the mountainside to release the water before it broke through the wall of ice.

CHAPTER SIX

FLOODING SEAS

Sea invasions

On a windless day, no place seems calmer than the sea. Yet when storms whip up its surface, the sea can surge ashore like a ferocious monster. Then farms, buildings, animals and people on low-lying coastal lands are at risk.

Sea floods can be especially severe where the coast is slowly sinking, as in the eastern United States and southeast England. But many scientists believe that sea flooding will happen on all low coasts, because the world ocean level is rising. These scientists argue that, by burning

If sea levels rise low island nations such as the Maldives could disappear.

fuels such as coal and oil, people world-wide add huge quantities of 'greenhouse gases' to the atmosphere. Carbon dioxide, methane and other greenhouse gases work like a greenhouse roof, trapping the Sun's heat near the Earth's surface. This extra heat might be enough to melt much of the ice sheets of the Arctic and Antarctica, and to release thousands of cubic kilometres of water into the world's oceans. Some calculations suggest that the level of the oceans will rise by up to 30 centimetres or more by 2050. Many major ports and cities could drown. Low island nations, such as the Maldives in the Indian Ocean, would vanish. Salt water would poison the supply of fresh water used by many cities and towns.

This chapter looks at some of the damage done by sea floods in the past, then explores how people have repaired the damage, and how they plan to cope with sea floods in the future.

North Sea storms

In 1928, a fierce storm in the North Sea sent a wall of water charging up the River Thames which flows through the heart of Britain's capital, London. The Thames overflowed its banks, flooding the Houses of Parliament. One elderly survivor later said:

'I was asleep on the ground level, when I was awakened by a terrific roar, and I stepped out of bed into knee-deep water. I rushed to open the door, but as soon as I did I was met by a wall of water which hurled it back. I struggled into the hall shoulder-high in water and clutched the bannister. I managed to get upstairs to safety when I heard terrible cries of "Save me, save me" from the basement. . . .'

Her basement lodger died.

Cattle are evacuated along a broken dyke in southern Holland after a night of gales in January, 1953. The Hondsbossche Barrier (inset) in the Netherlands is one of the many dykes built to protect reclaimed land from the sea.

Far more destructive were the sea floods that struck eastern Britain and the Netherlands on the night of 31 January 1953. That night the fiercest northerly gales ever recorded swept northern England and Scotland. Strong winds from the north drove millions of tonnes of sea water south through the North Sea faster than the water could escape through the narrow Straits of Dover. Waves piled up against the sea walls protecting the low shores of the Netherlands and eastern Britain. The level of the sea scarcely fell at all at low tide, and well before the next high tide, waves began breaking over and through the sea defences.

In Britain alone the sea swept inland in more than 1000 places. The Netherlands suffered even worse floods. For centuries Dutch engineers have reclaimed land from the sea by building massive banks, called dykes, to wall off huge sea-water lakes, and then pumping the water out with windmills. So, when the North Sea overwhelmed the dykes in 1953, water spread over an immense area.

All told, 1800 Dutch people drowned in the floods and 65,000 had to leave their homes, while 307 British people died and 24,000 British houses suffered damage. Even after the sea water had been drained away, the soil took up to 20 years to recover from the salt.

A disappearing coastline

For the last century the sea has made a slow advance against the low-lying eastern and south-eastern coasts of the United States. The danger is increasing, for as the sea level rises, these coasts are also sinking. By the 1990s, huge tracts of the east coast were disappearing at the rate of 30 centimetres a year. In parts of Virginia, the sea is advancing by several metres a year. On the coast of the Gulf of Mexico the sea invades even faster: Louisiana loses 40 hectares of marshy wetland every day. If the sea rose permanently by several metres, huge areas of southern Florida would be at risk.

Third World lands at risk

Apart from the Netherlands, the countries most threatened by a general rise in the level of the oceans are among some of the poorest in the world. One is Bangladesh. Much of the most fertile land of this overcrowded country is on a low, muddy delta. Here, the great Asian rivers, the Ganges and Brahmaputra, pour water sluggishly through many river mouths into the Bay of Bengal. Inundations of sea water have killed more people here than anywhere else on Earth. In 1970, 300,000 people perished when a cyclone over the Bay of Bengal drove sea water up and over the shore. In 1991, a cyclone hit an unprotected area of the Bangladeshi coastline and killed more than 100,000 people. Should the ocean level of the world rise by only 80 centimetres, one sixth of this small nation would gradually disappear under water. With the land would go more than 14 million domesticated animals, 100,000 small household 'factories', 8000 schools and 20,000 kilometres of roads.

Sinking cities

The famous northern Italian city of Venice stands just above sea level in a lagoon at the northern end of the Adriatic Sea, a long narrow arm of the Mediterranean. Venice was built on wooden piles driven into mud islands, and its main 'roads' are canals. For a long time the city has been slowly sinking in its mud, and since 1900 the seas have lapped ever higher. By the 1980s, sea water was flooding parts of Venice up to 40 times every year. Tokyo is another sinking city. This has been caused by pumping up huge quantities of underground water. Once the water is removed the land simply settles down into the space the water occupied. By the 1980s, more than 1100 square kilometres of Tokyo lay below the level of the sea, at risk from being over-whelmed by water driven inshore by the fierce tropical storms, typhoons.

Venice (right) is visited each year by over 12 million tourists but floods in St Mark's Square (inset) are becoming more frequent. Gigantic water gates will protect Tokyo (below) from the sea.

There are many types of sea defences. Wooden groynes and stone walls slow down incoming waves (main picture). Nylon bags filled with sand prevent the tide from pulling the beach out to sea (right inset). Marram grass is planted to anchor sand dunes (left inset).

Protecting the shore

For centuries people have protected low coasts from attack by the sea by building solid sea defences. Some are thick, high, concrete walls along the shore. Other defences include sloping banks of stone, rocks, bags filled with concrete and heavy, sloping metal grids. Three-pronged concrete blocks that interlock are used to break the force of the advancing waves. Stone or wooden walls, called groynes, jut out into the sea to protect some beaches from erosion.

After the disastrous floods of 1953, the Dutch and British greatly strengthened their sea defences. In the Netherlands a project was set up to build a network of dams and dykes to protect the vulnerable land against any repetition of such a storm. All the coastal inlets of the Rhine-Maas delta in the southern Netherlands were blocked off, except for the Westerschelde and the channel that links Rotterdam to the sea. The project took over 30 years to finish, but in 1986

the last flood barrier was completed across the mouth of the Oosterschelde estuary. Conservationists persuaded the Dutch government not to build a dam; instead a huge surge barrier was constructed which allows the tide to flow in and out, preserving the marine life in the estuary. Apart from monthly tests, the barrier is closed only once or twice a year, during threatening storms. The barrier is supported by 65 piers each weighing over 18,000 tonnes. In order to provide a solid foundation for these huge piers, the sea-bed was covered with 'mattresses' of sand and gravel held together by elastic fibres. These mattresses prevent the sandy sea floor being washed away. The construction of the Oosterschelde barrier took over ten years.

Unfortunately, solid sea defences do not always work. Defences that stop the sea overflowing on one stretch of shore are more likely to make the water inundate a less well-protected area of shoreline, and the sea may eat away at the beach below a protective sea wall or groyne.

The Sunny Isles Beach Restoration Project in Miami, Florida, USA replaced 25 kilometres of severely eroded coastline (left). The new beach (right) provides storm protection for the buildings along the coast as well as attracting tourists to the area.

For these reasons, engineers also studied the way in which Nature guards low-lying shores against the sea. In the USA and Britain, experts devised computer models of a coastline and its offshore tides and currents. They discovered that beaches, dunes and offshore islands helped to protect the land against high tides and waves by blunting the sea's attack. High seas may soak the coastal wetlands just inshore, but when the sea calms down again the flood water simply drains back out to sea. On many shores, marram grass has been planted to help anchor sand dunes, and to stop them being blown away. In Wales, fences have helped to build a beach by trapping windblown sand.

Water gates

Several countries have built gigantic water gates that allow ships in or out of ports and rivers when the sea is calm, but shut out the sea when a storm surge threatens to overwhelm the shore.

Dutch engineers have devised sluice gates to protect Europe's biggest port, Rotterdam. Dutch experts also helped to construct the Thames Barrier which has protected London since 1983. This storm-surge barrier is built on concrete islands across the river, like a row of giant's stepping-stones. Each island supports a curved steel pier, shaped like a gigantic gladiator's helmet. Inside the piers lie hydraulic mechanisms that operate four rotating water gates between the islands. With counterweights, each massive hollow gate weighs 3000 tonnes. When open, the gates lie on the river bed and allow sea-going ships to pass overhead. Locked upright, the gates create a wall rising nearly 16 metres above the river bed. Raising a gate takes only 15 minutes, and closing off the Thames completely, half an hour. This allows plenty of time for the gates' operators to close the barrier after they receive a storm surge warning from weather forecasters.

The Thames Barrier has just one drawback: its defences might not be enough. By 2050 the effects of sinking land and rising seas could mean that storm surges will overwhelm the gates.

After years of study, engineers also came up with a plan for saving Venice from the sea. Computer models of how breakwaters would affect waves and sediments showed where to place huge water gates at the entrance channels of the city's lagoon. These studies took account of changing currents, winds and waves, and the way in which the sea water would drag on the sea-bed and bounce off sea walls and the gates themselves. In calm weather the hollow, water-filled steel gates are designed to lie flat on the bottom of the sea across three entrances to the

lagoon, allowing ships of up to 140,000 tonnes to sail in overhead. When a storm surge threatens, air will be pumped in to drive water from the gates. They will pivot upwards on giant hinges to form artificial sloping beaches, each up to 350 metres long. The main problem could be that water shut out from the lagoon will overflow on to other less-protected areas of the Adriatic coast. The gates will be built if enough money is available.

Tokyo now also has huge gates for keeping out the sea (see page 38), and more are planned for the coastal city of St Petersburg in Russia.

Tomorrow's oceans

If the level of the oceans rises as much as many people predict, it will be impossible to save all low-lying coasts from being drowned. The main problem is the high cost of building sea defences. In the USA, for instance, the price of a sea wall only one kilometre long can be millions of dollars.

Even the richer nations could only afford to save their most valuable shorelines. In Germany the town of Hamburg is already spending the

The Thames Barrier (left) in London is designed to protect the city from flooding.

New varieties of crops such as tomatoes are being developed to grow on land that has been affected by the salt in sea water.

equivalent of two billion US dollars on flood defences. This is less than a quarter of the likely cost of the damage that bad floods would cause. Protecting England's vulnerable coasts could cost many times more.

High sea walls and river embankments might protect many ports, coastal factories and cities. Even so, salt water seeping underground is liable to damage city pipes and cables. Wealthy countries would simply replace them more often than they do today. But rebuilding drains and sewers as the water level rose would be a massive task.

Even wealthy governments are unlikely to spend huge sums saving low-lying farmland or coastal marshes. Governments might just have to let these areas slowly disappear beneath the waves. Already this is happening in the eastern USA. Instead of walling off tracts of lonely coast, the US Government has started paying householders large sums of money to have their wooden houses moved inland.

The problem of tackling rising seas is worst of all for poor countries such as Bangladesh. Another chapter describes grand plans to save this country from overflowing rivers (see pages 30-31). Yet Bangladesh simply cannot afford to pay for all the defences needed to guard against both sea and river floods. The same is true of other developing countries.

At least there might be a remedy for farm-lands poisoned by sea water. Scientists are developing food plants that could flourish where salty soil kills ordinary crops. American scientists have already crossed ordinary tom-atoes with a nasty-tasting but salt-resistant sea-shore tomato that grows wild on some Pacific islands. The result is a tasty tomato that grows well – even when the water sucked up by its roots is as briny as the sea. Another idea is to give ordinary crops the salt-resistant genes found in plants growing by the sea. If this proves possible, farmers might one day harvest salt-resistant asparagus, barley, rice and wheat.

But the best chance for saving the low-lying coastal areas of poor countries from the rising sea is for the world to burn less coal, oil and other fuels that give off greenhouse gases. This means persuading people everywhere to make more efficient use of fuel. So far this seems easier to say than to bring about.

Conclusion: preventing fires and floods

This book has shown what serious damage can be done when fire and water run amok, killing people and animals and destroying forests, farmland and buildings. Much of the damage to land and property can be repaired, but restored land and cities may still lie under threat. Limiting future destruction is just as important as repairing past damage.

Providing adequate early-warning systems can help in the fight against both floods and fires. Flood warnings allow people enough time to move out of the threatened area. Sprinkler systems and fire-extinguishers can be used to snuff out fires before they spread out of control.

Prevention is an even better way of limiting destruction. Many countries have laws which aim to prevent fires and floods from happening. Regulations may control the use of flammable materials in clothing, furniture and buildings. Strict rules help to reduce the risk of fire breaking out in aircraft, oil rigs, ships and trains. Flood waters can be controlled by the use of spillways, storage reservoirs, reinforced banks and giant water gates.

In many developed countries, modern technology has helped to curb the damage inflicted by fire and water. But sadly, major fires and devastating floods will always be a risk for those developing nations which are too poor to afford the protection that they need.

GLOSSARY

Amerindians – the native American Indian peoples of North, Central or South America.

asbestos – a soft, thread-like material that does not burn. It was used widely in fireproof clothes and to protect buildings against fire. Its use is now severely limited because asbestos dust causes serious inflammation of the lungs.

combustion – the process of burning. Burning usually occurs when oxygen atoms in the air join with carbon atoms in wood, coal, petroleum or some other fuel. Every burnable substance has a particular temperature at which it will combust. The lower the temperature, the easier the substance will catch fire.

conflagration – a large, destructive fire.

dyke – an embankment built to prevent flooding or to keep out the sea.

ecosystem – the structure of a system involving living things and their environment.

flammable – describes a substance that will catch fire easily.

girder – a strong, supporting cross-beam in a building. Steel or concrete girders bear extremely heavy loads.

hydrant – a water pipe with a spout and a valve. Fire fighters use street hydrants to obtain water from a mains water pipe running underground.

hydraulic – operated by water, or by pressure exerted by water or another liquid.

hydro-electric – electricity that is produced by the energy from flowing water.

inflammable – means the same as 'flammable' (see above).

lagoon – a shallow pool of water near to, or linked with, a larger area of water.

levee – a natural or artificial raised bank beside a river.

spillway – a channel that allows extra water to drain away from a dam or other obstruction without flooding the land around.